A Man and His Bees

The story of Alec Wilfred Gale

by D. A. Clements

Northern Bee Books

A MAN AND HIS BEES - THE STORY OF ALEC WILFRED GALE
© D. A. Clements

ISBN 978-1-908904-39-3

Published by Northern Bee Books, 2013
Scout Bottom Farm
Mytholmroyd
Hebden Bridge HX7 5JS (UK)

Design and artwork
D&P Design and Print
Worcestershire

Printed by Lightning Source, UK

A Man and His Bees

The story of Alec Wilfred Gale

by D. A. Clements

Northern Bee Books

CONTENTS

AN APPRECIATION

I first met Mr. A.W.. Gale in January 1925.

We were corresponding for some years before then, but that visit in 1925 formed the beginning of a lifelong close friendship. Up to the time of his marriage he spent several weeks each year at the Abbey, which gave us opportunity to discuss at length our mutual. beekeeping problems. After his marriage he continued to call now and again, but it fell upon me to call on him and in due course I had the privilege of spending many a memorable day at Church Farm up to the time of his death.

My own beekeeping goes back beyond that of

Mr. Gale when finally turned his energies to the production of honey. After initially trading in bees and Queens, our interests and problems were from then on virtually identical, apart from the fact that his beekeeping was primarily on an extensive and ours on an intensive scale. He gradually built up his colonies to close on two thousand, whereas ours never exceeded three hundred and twenty in number. On the other hand, our endeavours centred, apart from the production of honey, simultaneously on an all round improvement of the honey bee. To this end we at Buckfast kept a mating station in the heart of Dartmoor, accommodating more than five hundred nuclei. Our joint efforts proved mutually complementary.

The time we both took up beekeeping coincided with the period of transition from primitive to modern beekeeping in this country. The Isle of Wight epidemic, causing the loss of ninety per cent of colonies within the British Isles, took place at the same time. By the end of the First World War the old English native bee was gone. Likewise the keeping of bees in skeps. However, already before the demise of the native bee it was widely recognized that the imported varieties demanded a re-adaptation not only in management but also of hives and equipment. Anyone who did not actually witness the situation we were faced with at that time can hardly form an idea of the difficulties and problems we were up against. Beekeeping in this country had. clearly reached a turning point. However, dealing with the stark realities paved the way for the beekeeping of today. In the development of commercial beekeeping Mr. Gale assumed a leading role. As always his indomitable fortitude him through every difficulty.

Mr. Gale was undoubtedly an outstandingly competent beekeeper and apparently the most successful one here in the British Isles commercially. Needless to point out, he also possessed the necessary business ability, energy and stamina. The severely defective sight proved a lifelong handicap. However, an indomitable spirit and an unfailing optimism helped to disregard every adversity, difficulty and disappointment. Even when during the last years of life a serious deterioration in health and an almost total loss of sight manifested themselves, his cheerfulness proved imperturbable. As we know, beekeeping provided him with a full share

of calamities and failures, which were largely inevitable at the time in question. We were compelled to find a way through the maze of problems, on our own initiative, by means of failure and success. There were at this juncture no guidelines at our disposal indicating a solution to the many difficulties confronting us.

Modern beekeeping in any case lends itself to rash conclusions when not based on wide practical experience. Often a particular method is even nowadays put forward with great conviction and assurance, but when put to an objective test we are faced with failure. Two instances of this kind come to mind which caused Mr. Gale severe disappointments~ He experienced at one time some difficulties in. wintering his colonies. At that time special wintering cases, holding four colonies each, were in vogue in North America. He had forty of these cases made. However, we tried out this method of wintering in 1921 and 1922 and found that, whilst the colonies thus protected came through the winter in perfect condition and without a trace of mould on the combs, they failed to build up in spring. When Gale tried out this method of wintering we made use of the cases anew which we used five years previously. The tests were now carried out in two different parts of the country, involving more than one hundred and sixty colonies, the results merely confirmed the findings made in 1921 and 1922. One is tempted to conclude that our winters were not severe enough to warrant the extra protection. But in actual fact this form of winter protection was in due course also discarded in North America.

The other instance, though of a totally different kind, proved no less disappointing to Mr. Gale. We both attended a meeting of the Bee Fanners Association, where a new method of Queen introduction was advocated. Mr. Gale was highly impressed, whilst I could not bring myself to believe the method in question would work. When I met him the following winter he frankly admitted he had never before lost so many valuable Queens.

We freely shared our findings at all times. Whilst Mr. Gale proved outstandingly competent in many ways, mechanics were not his forte. The radial extractor, heather honey press and many of the labour saving devices and gadgets he installed or adopted were mostly made to our specifications. As to the kind of bee he favoured, he was at first convinced that a bee must be black to be any good.

At first he favoured the French black variety and later on a Caucasian and Carniolan strain and of course every type of mongrel that came his way. Admittedly each variety has some outstanding good points and mongrels usually manage to eke out an existence in seasons of failure without much attention. However, as Mr. Gale in due course discovered, a proven strain will over a series of years produce higher average yields with a minimum of care and attention. To his mind this decided the issue. In due course we also placed at his disposal some of our special importations.

This gave him an insight of the tasks we were engaged on and he thus rendered us an invaluable service. The comparative tests he carried out on our behalf have since his death been placed on an international basis. We were at all times not only interested in the satisfactory reports received, but possibly even more so in the unsatisfactory ones. These gave us an opportunity to eliminate promptly any undesirable line or cross.

Mr. Gale's findings were of special interest to us in regard to another aspect of beekeeping. His colonies were on British Standard combs, ours on M. D. size. The strain of bees was in both instances identical. However the differences in the yields of honey secured, determined by the size of the respective hives - or more correctly by the relative capacity of the brood chambers used - was remarkable, more especially on the heather. Here again, these comparative tests carried out over many years confirmed the findings we made before 1930. Mr.. Gale finally recognised the economic advantages of a large unrestricted brood chamber as provided by the Modified Dadant.

Whilst Mr. Gale was obviously an able businessman. He also was a most benevolent and generous person. Through suffering all his life ·from a poor sight, he nevertheless in his younger days visited the blind to read to them, whenever his time permitted. When I last met him, he confided to me that he would like to bequeath his entire beekeeping establishment to his men in recognition of the service they rendered him so faithfully over many years. The fear that in season's of failure they might experience some financial difficulties deterred him from carrying out this intention and eventually he adopted an alternative arrangement. As for myself, I owe him an everlasting debt of gratitude for the unfailing consideration he extended to me at all times.

His success in beekeeping was, apart from his unique business and organizing abilities, in great measure due to his untireing energy and readiness to devote every minute of his time to the needs of the bees when called for. Farmers must make hay while the sun shines. However, beekeeping on a commercial scale is a far more exacting undertaking. As I know from personal experience, no one can hope to make a success of beekeeping unless he is prepared to work hard and long hours in summertime. On the other hand, Mr. Gale's premature decline in health was possi. bly brought about in part by a disregard of his physical limitations. He had some hobbies, but these were set aside for the winter months.

To my mind his untimely death was a real loss to beekeeping in this country. He admittedly took little or no part in the meeting of beekeepers. In fact he obviously shunned the limelight and kept himself at all times in the background. On the other hand, his vast fund of knowledge at his command is something sadly lacking in beekeeping today.

Brother Adam

A W GALE - THE MAN AND THE BUSINESS

When I heard on HTV in Autumn 1982 that Gales were going out of business, I was rather surprised. It meant that they had kept going for thirteen years after A. W. Gale's death, which I suppose on reflection was very good.

The thought came to my mind - nobody has written the Gale story, and at that moment I determined to do it. Various things have happened to delay matters, but the time has finally come to make a start.

I worked there for about six years in the 1950's, so a lot of the material used and methods will be of that era.

My thanks to Mrs Gale, who has been a great help, and also to Allan Newman, who ran the business for the last thirteen years, and without whose help this book would not have been possible. Thanks also to Brother Adam for writing the forward, and was of course a friend of A. W. Gale for many many years.

Thank you also to daughter Victoria, without whose help I could not have managed to do this book.

D. A. Clements
Hilperton, Wiltshire 1986

ALEC WILFRED GALE

Alec Gale was born in 1900, the son of a Marlborough printer. The print works was quite small, having about six employees, so father was by no means a wealthy man. Young Alec always had ambitions to be a farmer, but money did not permit.

However he was sent to Marlborough Grammar School, and there passed a Scholarship for Oxford to study agriculture. There were only two each year, so this was quite an achievement. He passed this in July. 1915 - before even his fifteenth birthday - so it can be said with certainty that even then he was somewhat of a genius.

Before he got to Oxford however, along came World War 1, and by 1916 when he left school, all of his fathers employees had been called up. This meant helping in the print business until hostilities were over. In his spare time he helped with gardening and odd jobs for his father and the wives of employees away at war.

In June 1917 came probably the turning point of his life - he went out and got his very first swarm

One of the reasons for this was that food was short, and some honey would help out meagre rations. This swarm in fact produced sections that year, but then unfortunately succumbed to I. O. W. disease. By this time however, Alec had "Bee Fever" And I would think that every person who is reading this will understand that term., bee books become the order of the day, and he spent every available minute studying. At this time he also started to scour the vicinity for swarms, and to ask any beekeepers he could find to let him have any surplus ones.

By 1918 he had built up to about twelve stocks, but again he lost all but one with the dreaded I of W disease.

By this time he had I think almost certainly made up his mind to make bee keeping his living, although I. of. W. disease was then a serious threat, as there was no known cure.

'Then came the partnership with C. L. N. Pearson of Shalbourne, who had almost a dozen stocks.

The demand for bees at this time was enormous, so the idea then was to produce as many nucleus as possible and sell them. Italian Queens were the order of the day and were imported in great numbers.

In 1922 Alec bought out Mr. Pearson, and acquired all of the bees and equipment. For everyone who knew Alec, this would have been a foregone conclusion, as by nature he had to have his finger on the button.

He then started selling hives and appliances, and also selling Italian Queens in even greater numbers, as many as two thousand five hundred in a year. But the popularity of bees was on the wane and it was to honey production that he turned. His first real crop was in 1928, when he got about seven tons. This may seem a lot, and many commercial bee-keepers will never see that amount. But in later years when the business was really big, twenty tons was the breakeven point.

He rented some land in Marlborough just after this, but did not stay very long, for in 1931 he bought the land in London Road which right up to 1984 was to be known to everyone as the "Bee Farm".

In 1923 he married Miss Ferris, who was the daughter of a local auctioneer, and also at about this time he bought Minal Farm with a few acres When he eventually got possession, he set up home there with his new bride.

During this period the bees were being increased, and by 1936 he was up to almost one thousand six hundred stocks.

But farming was still a driving force, and in 1940 he bought Warren Farm which was at that time very cheap. The actual price I will not disclose.

Warren Farm was quite large, but this was not by any means the end of his farming ambitions.

Minal Farm was added to the list, and afterwards Whitechard, which over the years has been cleared of old trees and replanted, mostly with Norway Spruce. This area of forestry is about one hundred and thirty acres.

The total area of farmland was eventually about one thousand three hundred acres plus about four hundred acres in Wales.

In 1950, he took in Jack Ainslie as a junior partner to ease the work load.

In addition to this, there was an orchard at East Grafton of five thousand apple tree, Cox, Laxton and James Grieves.

There were three children - John born in 1939, who now manages the farm and two girls, Peggy (1934) and Jean (1936). Unfortunately, John was never interested in the Bee Farm and never took any part in that side of the operation.

Back to the bees again. Disaster struck in 1946, when EFB was discovered. This was almost certainly from the New Forest where they were taken then to the Heather.

This caused him in 1947 to destroy everything, and start from scratch with new combs and fumigated hives.

This was virtually the end of the sales of bees business, and from then on it was solely honey production and the sales of a limited number of Queens.

He bought two hundred and fifty imported nucleus, and in two years was back up to strength.

During the early 1950's~ he took over the stocks of A. D. Wequelin of Burford, which I think numbered about two hundred, so during my stay there the total stocks must have been around two thousand.

He unfortunately had very bad eyesight for most of his life, although he never complained. I can well remember we used to look over his shoulder at the combs to see the "Elephants", which was how the bees looked to us. For all that he could find Queens as well as any of us, although they were not marked.

At that time in the 50's we used to set off from the yard sometime before lunch, after Alec had done his tour of the farm and of the office work, which meant milk records and important letters.

He would come down the yard at a half run, shouting "Right ho chaps", and you were away on another adventure. Sometimes, if we were not to busy we would stop for half an hour, but at other times we ate between yards His lunch never varied, and consisted of cos lettuce mainly, of which he never seemed to tire.

Work continued usually until dark, and if the mood was good it was into a pub for a quick drink.

Mrs Gale tells me, that rarely was he home before midnight in the summer although she always waited up for him. We did not work on Sundays, but to him it was the same as any other day - looking round the farm, office work and anything else that needed doing.

Soon after I left, his health deteriorated badly, and at last he started to take thing a bit easier. Mainly, it was trout fishing and shooting that he enjoyed, although by this time his sight was very bad, and he could not dive himself.

Early in 1960 he had a bad coronary attack and from then on he got gradually worse.

To illustrate the immense courage he had, Mrs. Gale told me that even when almost blind he insisted on driving himself around the fame "Just see me over the main road and point me in the lane". Two days before his death in 1969 he was out with a shooting party - What courage.

THE SYSTEM

The Gale system I think was probably unique. In my own very humble opinion I think it was the best, although of course it has its critics.

The whole business was run on very precise lines, and everything was done in a very orderly fashion. Every gang had a specific job to do, and pressed on until it was done.'

I think the most important single factor was the "Gale" hive, which as far as I know was used only by Gales.

This was really at first-sight like a large National, but that was not quite true. The "Gale" brood chamber took thirteen British standard frames, which gave it much more breeding space, so that the first important fact was that every hive was always in a single brood chamber." This in itself gave it many advantages, one of which of course being that they were much easier to inspect, and of course were also much easier to load on to lorries, which was constantly being done.

The other major difference was that they were "Top" bee space, which although personally I prefer, could be questioned by many.

Excluders were of the wire type and were always wooden framed. This made them sturdy and rarely was a queen found 'upstairs'.

The next unusual feature was that virtually all supers were the same as the brood chambers, in fact I think that only about four hundred shallow supers were stocked. The other's of which about six thousand were kept were nearly all English Standard deep.. There were in fact about two hundred commercials, which were used in very heavy flows the combs in these were as used now in 'Commercial Nationals'. Incidentally, the record weight for one of these was one hundred and twenty-nine pounds.

As I see it, the systems big advantage was that all combs were· interchangeable, and this meant ease of operation.

Each stock had, for its first super a thirteen inch spaced body with about four or five foundations in and the rest fairly good combs. By good I mean without many drone cells.

Thereafter, any super that was added was spaced eleven combs - these always being second second class ones that had moved up - (explanation later). Metal ends were used in the brood chamber only, all of the supers had metal spacing strips which were cut both sides, so that they could space either eleven or thirteen frames.

Each hive had its own record card, which was normally used for two years, so that it could be turned back for reference to the previous year. Latterly all hives were numbered, which made identification that bit easier.

Entrance blocks were of a type similar to a National one, except that they had a metal strip guarding the entrance. These were only used in the winter, as when stocks were strong, "Summer Blocks" were used. These were just a long bar with a block of wood on either end to lift them so that it reduced the entrance height to about half, just to keep out any predator. This of course still allowed free access for the full length of the hive. Floors were of orthodox pattern, and entrance blocks were fixed tightly during winter by inserting a small wedge of wood at one end.

Crown boards had just an inspection hole in the middle,as all feeders used were galvanised ones that fitted right over the top - similar to Miller ones. These held about four gallons of syrup.

Hives were all placed on low stands - about three inches high, which were sloped slightly foward, so that the syrup ran to the front of the feeder. The advantage of these is obviously .stability, although it could be said that it was a bit more back-aching.

Tools of the trade were very. simple. We just used an ordinary screwdriver and a fairly wide flat scraper, about three inches wide. Veils were of the type that you had to use an old trilby with, but of course we did get through quite a lot one way and another. They were always " at the ready", but unless we were working in the rain or bad weather were very rarely worn. Of course there were moments when obviously they were needed for an upset or bad tempered lot. Gloves were unheard of, in fact I was so naive at that time that I did not know anyone used them.

When moving hives, shutting-in bars were used. They were merely very heavy perforated zinc fixed to a wooden bar, and these were fixed over the entrance after broods had been removed, with two nails which were partly driven in so that they could be removed easily. Screens were never used, in fact we did not have any.

Yards in the 50's always had forty stocks, so at this time there were about fifty yards which were scattered over a very large area, extending from north of Burford, to as far south as Andover. This meant that travelling was extensive and sometimes meant forty miles before you started work.

Latterly as forage got less, these were reduced to twenty per yard, and at one time there were ninety one of these.

Getting back to combs and to explain the previous statement of combs "moving up". An operation we called "CD" was carried out on every stock each year.

This meant Combs Done. Each hive's brood chamber was inspected, and any bad combs were put "above" and replaced by good drawn combs from the first super. Foundation was never put into the brood chambers. The vital task was to catch the Queen whilst looking through, so that combs could be moved around freely. She was placed in a match box when caught, and only released when all the work was finished.

Whilst talking of Queens, just to mention here that all were clipped - about one third was removed of one wing, but marking was never done, even in latter years. I was told that the reason was, that everyone was so good at "spotting" that marking was not necessary. Although perhaps boasting a bit, I must confess to still be fairly good at it.

Reverting,; to frames again, they were all the 'wired frame type', and were purchased, but again policy changed with rising prices, and they made their own, making perhaps one thousand or so each winter.

With foundation there was also a policy change due to expense. Wax was taken to Lees at one time ~and they made the foundation, but again eventually they made their own and only sold surplus wax.

Sites were selected by A. W. G. himself, and many factors were taken into considera tion. I suppose obviously the most important was the proximity of possible flora, but this became a more difficult problem as the years went by. For the majority of the period about which I write, there was no rape,so moves apart from Kent and the Moors rarely took place. In fact the only moves were to Rosebay Willow Herb, which when caught in the right mood was, and still is, a tremendous yielder.

Another very important consideration was accessibility. We had to get right in with a five ton lorry at any time of the year almost, but also to be away from too many children and any other curious people. Animals were not a problem, as all yards were fenced with two strands of barbed wire - we merely backed the lorry to one end and dropped the tailboard over the wire.   .

Regarding the proximity of water, this never ranked as a consideration and many of the sites must have been miles from any abundant supply. However no ill effects seemed to be suffered.

Another important factor was that they were not to far "out on a limb" as traveling was a major consideration. Once out on your daily round, usually you could be within ten miles of the next site, sometimes closer than that.

The rest of the system will hopefully unfold itself in the following chapters, which I will endeavour to make a little more interesting for the reader

SPRING

After being "confined to Barracks" for the winter, everyone looked forward to' the first outings, not thinking then of the hard long days ahead. At least five days a week until. dark, and quite a few "all nights" thrown in. However we were all anxious for the off, and waited to see who would get picked.

There were sixteen of us, and obviously at first only one or two gangs of three would go out.

The first job was floor changing, two hundred of these being taken out and stocks lifted whilst a clean floor was put down. This of course was an opportunity to see what food they had, and even more important that they had survived the rigours of winter. Losses were generally quite small, or relatively speaking so. Actual dead stocks were very few, but if you count weak ones, drone layers and those which for any other reason were no good, losses would be perhaps one hundred, which would probably be a bit less than the average.

An exception to this would be 1963 when the picture was a very sad one. This I think was quite general, and due probably to the duration of the hard weather, which prevented the bees from moving, hence they probably died in the midst of plenty.

This process of floor changing usually started about the first weeks in March; if the weather permitted. The whole operation took about a month, as each batch of floors had to be scrubbed in disinfectant and then dried. Each one was then inspected, and any necessary repairs carried out Obviously they had to be rushed through the workshops, as this was the only opportunity each year to put them back in good order.

When I said earlier on "to see what food they had", I meant purely by lifting to slide the new floor in, you would know the stores situation obviously they were not opened up.

Dead stocks would at this stage be loaded up and taken back. The combs in these which contained honey or syrup were very useful later on to use in Divs etc.

The weather now becomes slightly less important, as with April coming, Kent looms' ahead, and the rest of the programme has to be completed by then.

The next operation is critical, as the stocks selected for Kent have to be able to survive until Whitsun, as it is a very labour intensive operation to take feeders and syrup to the orchards.

So during the next few weeks every one is checked for stores and fed, if necessary. Also during this process, bodies are checked and any bad ones taken in for repair and creosoting~ Remember, it is vitally important that bodies are bee-proof as they have to be moved so many times. Also at this time, each one was given its first dose of 'Terrarycun' (more later).

A very interesting point is that very rarely if ever did I see Woodpecker damage. I can offer no explanation for this, unless creosote does deter to a certain extent.

The selection of Kent stocks follows on from this, and also all record cards are turned over (or changed. Only the very strongest were selected, as it was felt that the better the service the more chance there would be of repeat orders in the coming years. Once these tasks had been completed, the Kent stocks were nailed up (floors to bodies with two double-pointed nails each side). This could be very unpleasant at times, as everyone knows, bees do not particularly like the music created by hammers, so we worked in threes - one each side with nails and hammers and some swift puffs of smoke from the 'smoker boy'. The secret of this seemed to be speed do it quickly and move on. Having by now taken feeders off, the end of April approaches and the pressure is off for a short time.

Grass is growing by this time in the yards, so the annual 'purge' can begin. The treatment was at this particular time always creosote - it was very cheap, and a few hundred gallons were used every year.

A forty gallon drum was loaded up each day and water cans used to sprinkle it round each hive. About six to nine inches round each one was cleared, and
the stand given a liberal dose at the same time.

Grommoxone was ultimately substituted as creosote became ever more expensive, and I suppose in all honesty it is much less messy to use and probably
more efficient. I must just mention here one other 'painful task'.

As I have just said, creosote was used six to nine inches round the hives, but what of the rest of the grass and weeds? There was a primitive solution to this problem, it was known as the 'Allen Scythe'. Bees seem to hate two stroke engines to begin with, but if you imagine trying to manoeuvre one of these monsters between the hives and round the turns at the ends, it was quite a performance I can assure you. As you will know if you can remember them, the blades protruded at either end of the cutting bar. Well the engine was one aggravation, but the rat-tat- tat of the cutting bar on a hive if you misjudged was an additional one. And to make matters worse it was almost impossible to stop them when you wanted to. If you can imagine half an hour of this treatment, you can be assured that "temporary withdrawal" was not unknown.

This was very definately a period of long hours, but being a lorry driver meant variation of location for me. In fact I found it very exciting.

The signal that things were on the move in the orchards was when Walter Smith, our foreman, brought the caravan from its winter quarters and gave it a spruce-up. Just a matter then of waiting for the phone call. to say that the blossom was at a stage when the bees would be needed in the next few days.

Walter then hitched up to one of the lorries,and taking just one boy with him was on his way. They set up the caravan at Lady Dane Farm at Faversham, which a fairly central point for them was, and there it remained until just before Whitsun. Usually the first loads went down in the second or third weeks of April, but of course so much depended on the weather.

On the day that loading commenced, the person who was to drive down at night usually went home early, but this depended on the work that had to be done.

Usually the loading gang went out with one lorry but if yards happened to be very far apart, two might of go out .and then transfer to one back at the Bee Farm. Each vehicle was equipped with high sides, about six feet above the bed. They were not solid, as the timbers were far enough apart to provide plenty of air space. These were usually left on the front half of the year; and with a tarpaulin over the top,provided a dry area for men and equipment. The only modification when carrying bees was to take out the bottom board at the front to provide a through draught.

On arriving at the loading site, all of the entrance blocks were pulled out and collected up, this was to avoid disturbance later on. A shutting-in bar was thrown down by each hive. As soon as most of the bees were in, the shutting-in commenced, sometimes leaving a few behind, as if you waited long it was dark long before the end of the job. They were loaded three across, six high and eight rows making a total of one hundred and forty-four.

Travelling was always done during the night, so the driver left the Bee Farm usually about midnight.

The journey was quite long, depending on which way you went, it was between one hundred and twenty and one hundred and forty miles. You would meet up with the gang who were living in the caravan at a pre-arranged orchard at first light. Normally a tractor and trailer was provided by the grower, as it was often impossible to get between the smaller trees with the lorry. After unloading as quickly as possible it was then away to the nearest cafe for a well earned breakfast. After an hour or so it was back at the wheel and home to Marlborough. Tackographs and log-books were fortunately unheard of then.

This process continued until all were safely in Kent. Usually the total number was between one thousand and one thousand two hundred. The area that was covered there was quite large, extending from north of Maidstone down to Faversham, Yalding, Canterbury and Ash.

A gang of about. three stayed in Kent for the duration of the blossom, as there was quite a· lot of work to be done. Mostly this entailed moving stocks from cherry to apple blossom, but also to any other type of fruit which needed pollination. Also they had to be checked and if necessary, given supers. If these were necessary a special excluder was used.

This basically was a crown board with about a six inch square excluder in the middle. In this way if the weather got very cold it prevented them getting chilled.

Some of the growers remained loyal for over thirty years,which proves that if the service was good they were happy.

Basically the return home was a reversal of the process, but at times it could prove more of a problem. Usually not much honey was gathered in Kent, but in one year, which was 1978 there was a huge crop. 'This created many problems, and slowed down the return process.

If caught with a very mild spell of weather, Guilford fire station was a stopping place, and they obliged with a good long hose down, but generally there were no such problems.

It was reckoned to get all stocks home by Whitsun, which is the last weeks in May or first in June. They were all sited in the yards then, in which they spent the summer.

Of course in the early days there was no Rape, but a large commercial bee keeper would now face a dilemma - to go for the certain money. of the orchards or to hope for the very much bigger returns that rape can give. I would personally think that the prudent man might put.some eggs in each basket. But having said that, we all know that rape can be dismal failure whereas the orchards never let you down.

BACK TO THE YARDS

After the humdrum few weeks at Kent, things settled down to a more sedate pace - still very long hours but a more regular pattern of life.

The big long job that loomed was the "CD". As I explained earlier this was very time consuming, in fact the labour involved was such that today it would be uneconomic. Every single stock was done every year, and varying numbers of combs taken from the brood chamber. Of course doing it annually no combs got terrible, but as many as five or six were sometimes moved up.

Whilst this as being done, supers were given, if required and obviously cells broken down. Every one was destroyed at this time but a policy change meant that one was left in every case. As queens were all clipped it was obvious if a stock had superseded, so no undue problem was created.

During this period the ten days inspections went on as usual or in a very heavy flow they were obviously supered whenever necessary. Sections were tried only on very rare occasions. The success rate was not good enough to make them commercial viable, although they seemed in great demand.

Cut comb was sold on a fairly small scale, as labour was not available to produce it.

Swarms were a very rare sight which does sound almost impossible, but I don't think I actually remember seeing one.

Really there is not much more to say about this particular period of the year. It is in the next period that the hard graft really begins.

During this summer part of the season Alec' always did the driving himself - nobody else could get there quickly enough. Lots of the roads to the sites, although perhaps B-class ones were very narrow. The Boss never slowed up - thinking he had some divine preferential right. Many times cars have hastily got up the banks to safety tooting their horns furiously.

Alec would often pass some comment to the person in the cab with him such as, "What's the matter with that silly idiot?" All in all considering his eyesight he escaped very lightly and as far as I can remember never had an accident. This can be attributed mostly to the evasive skills of other drivers.

EXTRACTING

This had to get under way as soon as possible when the flow had finished, as extracting itself was an enormous task. It would normally involve about five people directly.

Once the first lot of clearers had been put on, it was then only a matter of a couple of days before things got under way.

There were two radial three-phase extractors each taking forty-four British Standard deep combs. Usually the supers were taken off late in the evening to keep robbing to a minimum, but with forty stocks in a yard it was normally impossible to avoid it completely. Clearer boards were left in-situ at one or two yards that were fairly close to Marlborough. These were the stocks to be used for drying supers, as none were ever stored wet.

Supers being unloaded were put as near to the extractor as possible, and as they were completed were stacked the other side of the room here they could be loaded to be taken out for drying.

It was almost a conveyer belt system, which in very good years did not move fast enough to keep everyone happy.

The uncapper itself was very efficient, and the poor chap operating it stood in one spot pushing combs over it for weeks. The knife was horizontal, and was driven by ,an electric motor, and moved very quickly backwards and forwards. Experience taught you to keep fingers out of the way. The blade was hollow, and attached to a steam boiler by a length of rubber pipe, so it was constantly very hot. The combs were just pushed down over the blade - holding tightly to the lug on one side of the frame. You then just pulled it back up, turned it over, and let it go over the knife again. The comb was then placed in a rack, and they were loaded in the extractors as quickly as possible.

The cappings dropped down into a large tray which contained an heated element. This heated very quickly, and from the bottom draw-off valve you could take the honey out and pour it into the extractor. The wax could be taken off the top when necessary. Each extractor was connected to an electric pump which pushed the honey into a huge tank in the next room. I am not sure what the exact capacity of the tank was, but probably two or three tons.

When this was fairly full and had settled down, the job of filling twenty-eight pound tins commenced. These were carried into the adjoining room in which the cellar was situated. They were then lowered by rope and stacked in a separate place according to their year. It was always an objective to have a carryover of stock so that customers could always be supplied. No honey was ever sold in bulk.

To get to the cellar was always pleasant, as in the corner a large barrel of mead was always kept on the go, and this was an opportunity to sample it.

The taking out of wet supers continued throughout the extracting process. On arriving at the yards the clearers were removed exposing excluders, and eight supers were put on each hive. This operation was carried out late in the evening, as robbing was a terrible problem. The job was done with all possible speed and supers on the lorry were covered by a tarpaulin until the last possible moment. If robbing became out of control you had to pullout and take the remainder of supers to another. yard.

As soon as they were cleaned out, clearers were put on and supers were stacked in the store. Those with thirteen comb spaces were put away for attention later.

The eleven spaced ones were put ready for use on heather stacks.

EXMOOR AND DARTMOOR

The moors operation started in 1947, and Gales took bees there every year until they closed down. Usually about six hundred stocks were taken, these being mostly put on Exmoor, in the area around Exford and Wheddon Cross. A few hundred were taken to the Mary Tavy area of Dartmoor. This would depend on the state of the heather. Taking them down also depended very much on the weather but the date would usually be somewhere in early August. The procedure followed very much the pattern of the Ken t move but with some variations ..

The first and most important thing was that only the six hundred very strongest stocks were taken. Weaker ones are no good at all, and much better off being left at home. Having selected the ones to be moved, each one was given an eleven spaced super. A full comb of honey was placed in each one as an insurance against very bad weather. This of course would not keep them long, but it gave them that bit of help. These supers always contained the very oldest combs available. This was as it were' the end of the line' for them, having by then worked their way right through the system, they had certainly paid for themselves.

Supers were fixed when put on with double-pointed nails, and floors were checked to see that the nails were still secure.

Close contact was kept with the farmers on the Moors, and as soon as the ling was almost ready, the caravan was on its way again. It was parked about a mile from Wheddon Cross, and stayed there (as for Kent) until the last hive was on its way home. The loads were obviously less than for Kent and seventy two were taken at a .time.

Loading and unloading followed the same pattern, except that they had to be carried on the Moors as some of the sites were not very level or accessible.

Heather as you know is a very chancy crop depending very much on the right weather at precisely the correct moment. If everything is right the flow is cnormous, but only over a fairly short period of time. To me, just the wonderful aroma that prevades the air makes the whole thing worthwhile.

However, very rarely is an additional super ever needed.

When the flow was over, and this again was very dependant on conditions, clearers were put. on and every super removed before transportation. There was a very good reason for this, mainly that it speeded up the process at the Marlborough end and enabled feeding to commence immediately.

In the days up to 1953, supers were taken straight to Buckfast Abbey, where Brother Adam kindly allowed us to use his very modern press. I think at this time it was probably the only one in" the country.

Often the weather on the moors was awful, and rain and swirling mists could be very frightening. I personally was quite afraid of getting totally lost on those narrow winding roads and always happy to be safely back to base. One night that I will never forget was on Exmoor. We had completed loading and roping down but in trying to get back onto the road the lorry almost turned over, and the load slipped.

We spent the next two or three hours, in absolutely pouring rain, transferring the lot (one hundred and forty-four) onto another lorry, after which I had to face the drive home. We were all absolutely wet through, and certainly would not want to go through that ordeal again.

During the summer months we often had students to help out, and to get themselves some experience. One such lad was Lionel, who hailed from somewhere in Yorkshire. He was a nice lad, but insisted that bee stings were just like pin pricks, and that no matter how many he got it would be "No bother".

As most of you will be aware, bees can become unpleasant whilst working the ling, so care is always taken.

However one night we were loading up and Lionel was helping us. Things went very well until our good friend Lionel tripped over a boulder. The rest is very easy to imagine, but just, to say that he was evidently very poorly' afterwards. We were never able to Question him however, as he as never seen at the Bee Farm again.

In about 1954, Gales purchased their own Heather press, which was a similar machine to the one at Buckfast. It took about forty British standard deep combs, and exerted a pressure of about twenty tons per square inch. This machine indeed made a tiresome task very much easier. All heather honey was sterilised before being stored way in the usual twenty-eight pound tins.

Rent for sites was paid by way of a comb of clover honey. It was strange but it was invariably preferred to their own heather comb.

The major objective of going to the Moors was of course the honey, for which Gales had a very large market. This was mostly abroad. In fact the majority at that time went to America.

The other important factor was the lessening of the amount of feeding. that had to be done, meaning a great saving in the sugar bilJ.

If the brood chambers were solid with heather honey, two or three combs were removed and replaced with empty ones and a small. feed given. This was to reduce the risks of dysentery, which seems to break out when wintering entirely on heather honey.

FEEDING

This was a very messy job to say the least, and at that time done in a very antiquated way, specially as far as making the syrup was concerned. Whilst the remainder were on the Moors, the thousand or so left behind were fed and wintered down if possible. Obviously there were not enough feeders to go round the whole lot, so about five hundred at a time had to be attended to.

Syrup was usually made with ordinary sugar, but one year partly refined Barbados was used and seemed to be very successful. An old copper was used for the mixing, and this was heated by logs, so as you can imagine the quantity made by each boil-up was not very large. One person had a full time job for a few weeks and apart from the heat and stickiness, wasps were. an ever present threat. The constant smell of syrup attracted vast numbers, and it was 'very difficult to avoid getting stung.

Thymol was added to all winter feed, although its merits were under constant debate.

As each copper full was ready it was ladled out and put into four gallon jerry cans They were loaded onto the lorry and had to be carried to each stock in the yards. Two cans were very heavy, and it was always a very tiring job.

Each stock was now fitted with a Winter Block, and this was fixed firmly by inserting a small .wedge at one end.

A. W. G. himself always supervised the feeding and seemed absolutely indefatigable. He lifted every one and wrote on the record card the amount to be fed. Someone came behind and put the feeder on and filled it to the level instructed. The feeders held about four gallons so usually unless they were particularly light, one feed was sufficient for most. When all available feeders were used, a start was made again with the first yards to be done. These were again lifted, and if heavy enough the feeders were removed.

When the crown board was replaced, a chip of wood was placed under each rear corner, making a gap just less than a bee space. This provided enough ventilation to prevent excessive condensation.

During the feeding procedure, the double-pointed nails were removed from all of the hives. This of course was necessary for the. early season floor changing.

Yards were consolidated for the winter and even in later years when only twenty were in each place they were doubled up to cut down on the. traveling.

By the time the Exmoor stocks arrived home, if things had gone to plan, feeders could be put on as· they were unloaded. Even working quickly, by the time that six hundred were fed, the month of September was nearing its end.

Actually, Walter always held the opinion the later you fed the better, although of course with huge numbers to contend with, you could not afford to wait in case the weather suddenly turned against you.

I recall going to my first Local Beekeepers meeting and hearing tails of "Mouse Guards". Not wanting to feel a fool I said nothing, but discreetly found out what they were afterwards. I mention this just to illustrate one of the very many differences between a few stocks in the garden and the world of commercial beekeeping.

Incidentally, instances of mice entering hives were very rare. The metal slotted plate on the entrance block meant that there was no way in - apart from eating through about three quarters of an inch of wood.

Once the feeding was finished and all the feeders removed, they were not visited again until the floor changing routine the following March.

QUEEN REARING

I must now tackle what is probably the most important facet of all. Successful bee keeping must hinge on a good Queen rearing policy.

The system I shall explain is the one Gales used with few modifications for over fifty years. This does not however mean that it is the best way for everyone, or that others .will entirely agree with it. It was the way however that best suited their purpose.

I will record here that as far as I know, Brother Adam sent one Queen every year as a breeder, and therefore a high proportion of Gales strains were based on Buckfast ones. Another major advantage of course was the tremendous choice available. Breeder stocks were carefully selected the previous year, and care was taken not to be influenced too much by an excellent stock that was. on the end of any row. This was in case there had been excessive drifting into one so placed. Many other factors were taken into consideration, which included temperament. Also the ability to get a good surplus when available, and if possible, ones that did not need excessive amounts of syrup to feed down for winter.

When the twenty-five or fifty stocks needed had been sorted out, they were then marked. A few extra ones were added in to cover any losses that might occur before the spring.

When loading for Kent they were picked up and taken to the Queen yards. Ten were put in each of the large ones and five in the smaller one. About a month prior to the commencement of operations they were given feeds to boost the strength as much as possible. The two larger rearing units had about one hundred and ninety div boxes in each and the smaller one about ninety.

The first grafting of the season took place around the third week in May. If things were advanced and the weather good this could .be considerably earlier; though NEVER before May 8th Work commenced at each of the Queen yards at about the same time, although obviously grafting was staggered somewhat to avoid the problem of having more work than could be coped wi h at anyone time.

Grafting was done in the Queen yard concerned, and a nice day would be done sitting on the step of the shed. If it was thought to be not warm enough, the cab of the lorry was used. Grafting was done into wax cups which were made at the Bee Farm. These were fitted into a round wooden base. They were then attached to a wooden bar which was slotted into a British Standard deep frame. Each bar had eighteen cell cups attached to it. New drawn comb was put into the breeders stocks at the appropriate time if possible, so that these could be used for grafting from. One of the rules that applied whilst this was taking place was "no smoking". All grafts were ALWAYS done dry. The success rate was very high, and often sixteen or seventeen or even all of them were usable cells.

The stocks to be used for the rearing were all de-queened, and used for as long as possible. By this, meaning until they became tired of the work and then another one would be prepared. About eight or nine stocks were used during the course of a season, one in each yard being kept going all the time.

Strength was kept up by introducing brood regularly, which kept the nurse bee population high. They were given a feed of syrup regularly, so that there was never a food shortage.

On the first occasion they were used, each stock was given four bars (seventy-two cells) but as they lost zest this was reduced to perhaps two bars, and ultimately they would be replaced by a new or recently de-queened hive.

During the late afternoon of the day they were due to hatch, a large gang departed on a splitting mission. This would take place at any convenient yard that was not too far away. Three people would go in front and catch the queens and put them into a match box, leaving them on top of the frames. The gang behind would very quickly split them down into about eight divs from each hive on average. This would consist of one comb brood, one comb store and one cover comb. They were equalled out so that each div had about one and a half combs of bees. These div boxes were then loaded onto the lorry for transportation.

Another gang would them make up the number of combs remaining in the hive to eleven. Into the remaining space was put a-side feeder which was filled with syrup, and then the Queen was released. With any luck they would build up enough to be ready for a late flow or for the heather.

The number split depended obviously on the number of cells that had taken on the batch that were ready. This meant that if for instance one hundred cells were ready, you had to split about twelve stocks. The divs were then transported to the Queen yard they were needed at and put onto stands. They were left for two hours before being given a cell. This was to make sure that they were receptive, and not liable to tear the cell down.

Divs were only covered by a piece of cloth under the roof, this was to prevent robbing, and to enable you to look in as quickly as possible and close them up again.

Entrances on most of them were kept high. This was for a very good reason. Feeding all divs was by means of dry ordinary granulated sugar. If the entrances are on the bottom they will remove it by pushing it outside. By having a high entrance this is impossible, so they happily eat it. After mating, Queens were left for about ten days until brood about the size of an old Half Crown could be seen. Although no actual guarantee, it usually meant the Queen would be all right.

The process of requeening then began. First attention was given to weak stocks, drone layers or any queenless ones. These had all been marked during the normal routine inspections.

Introduction was by means of. a normal two apartment cage. One end was filled with candy and the other end was covered by a zinc gauge to prevent release from that end.

Candy was made from just honey and icing sugar, but care needed to be taken not to make it too hard, thereby trapping the Queen for an excessive period of time. A two inch nail was used to pierce a hole right through the candy to make it easier and quicker to get through.

On arriving at a yard, the ,first priority was to catch the old Queens to be replaced. These were then immediately killed by chloroform and sold for research purposes.

The cage was inserted straight away and placed over the brood nest. They were then left entirely alone for about eight days and then checked. If they had been accepted, the cage was recovered. In cases of failure, cells if' any were broken down and another Queen given immediately.

The acceptance rate was very good normally, and out of say sixty introductions it was normal perhaps to lose three or four. This could at times rise to seven or eight if there were adverse circumstances.

Having completed the first batch, the process carried on with the rest until all that needed replacing had been done. I must mention again here that all Queens were clipped, so in the case of finding one not done in a stock it was counted as being a year older than was known and replaced after one year. Normal good healthy Queens were allowed two seasons.

Increase

When all of the requeening was finished, the making of annual increase was the next job. In a normal year about. one hundred or one hundred and fifty would be needed. This number would replace winter losses and other unforseen happenings during the season.

The method was quite straight forward, and followed to an extent the earlier method. In this case, two combs of brood were given to each other and added to this would be one of food and two of Cover combs. If Queens were available they were used. A side feeder of syrup was then given. The objective was to get them up to seven comb strength for winter. The golden rule was that all increase had to be made not later than July 1st.

With this operation completed, the Queen rearing yards were closed down for winter. Cell rearing stocks were requeened and all divs were taken back out to yards and united up with flour.

I must express my very sincere thanks to Ron Lane for all the help he has given me generally, but particularly for his assistance with this section on Queen Rearing.

WINTER

This at least was the time when the pressure was off. You would imagine that to find jobs for sixteen people from September to March would be very difficult. In fact this was far from being true, and unless the weather was very bad, everyone was well and truly occupied with one or another of the many tasks.

Firstly I will deal with the work actually done on the Bee Farm. Usually two people spent most of the winter working on bodies, floors, crown boards, clearer boards etc. All of these items were checked carefully and any repairs carried out. In the case of clearers, all of the porter escapes were cleaned up and made to work freely. All, spare roofs were also inspected and repaired.

Some unlucky person was given a chopper and hundreds of sacks to mutilate. you can imagine how much smoker fuel was used in the course of a season:

Smokers were dismantled and cleaned out, and if necessary bellows were given a new leather cover.

Then there was a pile of about two thousand excluders to be dealt with. Each one was carefully cleaned, and had a visual check to see that all wires were parallel. Excluders were always treated and handled very carefully, as so much depends on their efficiency in the season.

If the Heather crop had been good, hundreds of frames had to be cleaned off and rewired (all were of the wired frame type).

Another end product of this was a mountain of old pressed combs. These kept another person busy for a very long time with the wax extractor. This machine was very efficient, but even so it was a very laborious process.

The massive job of sorting out combs then had to be done. Everyone was inspected, and old combs were put into the eleven spaced supers. The good ones remaincd in the thirteen spaced supers, which of course received all of the foundation. Usually, six to eight foundation were put into each super, but this could vary immensely, depending entirely on the Heather crop and the number of other combs broken during the season.

 Fitting the foundation was another lengthy job, and this carried on throughout the winter, as with so many to do, it could not be left until spring.

Bodies were regularly creosoted, and any necessary painting and repairs done to the buildings. The four lorries were given services and painting or other repairs done to them.

 Apples which had been picked in the Autumn were sorted out and packed for sale. These were mainly Cox's, but Laxton and James Grieve were also grown.

Bottling and labelling of honey was done fairly regularly, as normally large stocks were not kept in jars.

At that time major customers included Boots, Fortnum & Mason, Civil Service Stores and many other national concerns. This was all delivered by lorry. Boots went to their main warehouses - one in London and the other in Manchester. Whilst in the Midlands it was usual to go on to Knottingly in Yorkshire to collect a load of one pound jars. London deliveries were usually of two days duration, and perhaps as much as five tons may have been distributed around the capital.

There was often a load of sugar to bring back from the docks - a very unpleasant job. It was usually 3cwt sacks, which were loaded by crane from an upstairs warehouse, five or six sacks bundled together. Standing underneath it whilst trying to sort out the tangle on your own was to say the least a very sticky affair, as usually the bags were leaking. Then followed an unpleasant seventy miles drive home, with the thought of a bath and change of clothes uppermost in your mind.

For those of us that could be spared from inside jobs, there was always plenty to do on the farm and in the woods. A number of buildings were erected by Bee Farm staff, and in the woods there was clearing and tree planting to do. Also, in December large numbers of Christmas trees had to be dug up and delivered to wholesale customers.

DISEASE

This subject I have left purposely until almost last. It is never a pleasant aspect of bee keeping but one that has to be faced.

Gales in common with all other commercial concerns had their problems over the years. What I think most people fail to realise is that when working on this huge scale it is absolutely impossible to keep equipment separate in any way at all. If for example the national average of diseased stocks is one per cent, this would mean that with ninety-one sites you would be almost certain to get it in one. Having got even one infected, the problem multiplies. Apart from equipment problems, there is also the added factor of having so many moves in a year. Firstly Kent, then possibly to Rape, some to Queen rearing yards and then to Exmoor. All of these stocks obviously get mixed up each time they are moved.

This brings me to the subjec which causes more controversy than most - Terramycin Powder. I understand that Alec Gale was one of the first users in this country, having got some from America. The fact that it very effectively suppresses E. F. B. is undeniable. From reading the available literature it seems that it is standard treatment in most countries and is used annually as a matter of routine.

I have personally talked to a number of professional beekeepers, and the consensus' of opinion is that it is very difficult to cope without using it. The remarkable effect it has on clearing up E. F. B. symtoms is undisputable, and this I can personally vouch for.

One answer for a smaller scale bee keeper could be in the Gamma Radiation Plant at Swindon. The cost of this though would be utterly prohibitive for doing say eight thousand supers (£4 - £5 each).

Contrary to the opinion of some people, Gales did not close down because of disease. Basically they managed to carry on longer than A. W. G. had hoped, and eventually felt that an undertaking on this scale was difficult to keep viable in present day conditions.

For various reasons, I do not intend to add anything more to this chapter, feeling that there is nothing to be said that would be. constructive or help any cause.

FACTS

I have collected together a few interesting figures, some of which are quite amazing.

Unfortunately lots of papers were destroyed when Gales moved offices, but fortunately the worst and best crops and years have survived.

1934 The worst year ever, Only about ten tons from nine hundred stocks.

1935 By this time the business employed eight full time staff.

Alec Gale took Gales of Peterborough to court to try and stop them trading under the 'Gale' name. Unfortunately he lost as they had in fact used the name first.

WAX In a good year there could be one ton of wax for sale after making their own foundation.

SUGAR In one season, which was a bad one, forty-five tons of sugar was used for winter feeding.

HEATHER The best crop in any season was twelve tons.

YIELDS The best years were as follows:-

1955 .... 89 tons
1959 .... 86 tons
1961 .... 88 tons

1978 Very much against normal trends Kent produced an extraordinary crop of approximately forty tons, making it difficult to get stocks home.

Honey was exported to America, Australia, Canada, France Sweden, Italy and Cuba.

It was reckoned that a twenty ton crop would just enable Gales to break even financially on a season.

Average super full was sixty pounds of honey.

The record gross weight of a Commercial super was one hundred and twenty_ nine pounds.

The winter weight of a hive when fed was about seventy pounds.

QUEENS About eight hundred had io be produced each year for their own requirement.

STAFF The maximum was about sixteen but this number did decrease somewhat as overheads continually increased.

Some of the employees were very long serving and in fact even after retirement continued to work part time.

WALTER SMITH - Foreman for many years. Sadly he died in 1983. Continuously employed for fifty two years.

ERNIE GIDDINGS - Thirty-five years on Bee Farm fifteen years in gardens.

Completed fifty years in March 1985.

JACK MADGWICK - Thirty-five years with the bees.

REG PAYNE - Thirty years with the bees.

RON LANE Started in 1957. Foreman after Walter retired from full time work - still employed.

ALLAN NEWAN - Part time and full time since 1953. Still secretary now and was a director of the Bee Farm.

These must point to the fact that A. W. G. was a good employer, and every one of the people who worked for him for any length of time were very loyal.

CONCLUSIONS

By sheer chance I was reading this week from a bee magazine, the following, "His target is five hundred hives by 1986/7 producing thirty thousand pounds of honey in a season".

Unfortunately this sort of statement bears no resemblance whatsoever to reality. If it were only partially true, Gales would no doubt still be in business. Maybe perhaps even the concerns smaller than Gales, like Manleys would have found a purchaser or a way to keep going.

To reiterate some earlier points, basically the vital ingredient~ are now missing. Namely that the removing of hedgerows, the loss of permanent pastures, and the use of pesticides and herbicides have all reduced the possibilities of success. Add to this the zooming price of petrol, and wages which have risen enormously and the problems begin to pile up. The ever-present natural hazard of the weather and very high sugar prices only make matters worse. And of course the price of honey has not really risen enough to compensate for all these things. A quick calculation shows me that to keep solvent, based entirely on bee-keeping, Gales would need about sixty tons per annum. Economics in staff however drastic, do not allow the gap to be closed far enough. And added to this, standards would inevitably have to be lowered.

Looking now to the bright side, there are ways of getting a financial return from a limited number of stocks.

The first priority in my view is that the rape crop is an absolute essential. Of course some people .may not be near enough to be able to get stocks to

it without crippling expense. In this case, pollination is still lucrative if again travelling is not too crippling a factor.

For both of these operations, quick and efficient spring build up is absolutely essential. The answer to this is in my view, not so much in spring feeding, but in getting them wintered down with sufficient stores to cope with any eventuality.

Many many people fail to do so properly, sometimes perhaps by genuine mistake, or perhaps by trying to save a few pounds of sugar.

Going to heather is again an important factor. It does not always payoff by way of crop, but even so usually there is something for the brood chamber to help reduce the sugar bill. The more you can take obviously help the transportation costs per hive.

In the spring of last year I was offered ten complete stocks in Gale hives. The vendor had severe back troubles and by necessity had to drastically reduce his workload. However his loss was my gain. I bought with each one a feeder, excluder and two supers, but sadly without any good combs. This meant working last season on nearly all foundation in the .supers.

Fortunately, the rape turned out tolerably well in the end. For the first week or two it was in flower the ground was rock hard, but in the third week there was a terrific storm, and the flow commenced in earnest. The outcome was just over four hundred and fifty pounds in weight, which was very satisfying. From then on the weather here like almost everywhere was hot and dry making the normal summer crop very disappointing. In fact this produced eventually about two hundred and fifty pounds in weight - which included the cut comb that I did.

In early August the ten were on their way to Exmoor. Fortunately, a kindly farmer remembered me from the Gale days and took pity on me, as sites on the moor are not easily found these days. Looking back on it now, the feeling is that I was a few days late and probably missed a little of the main flow.

The reasons for this were twofold. Firstly it was difficult to find the time and secondly there were things like Willow Herb still in flower. Obviously getting there too early could mean a mixture in the supers, and this I wanted to avoid. The outcome was fairly satisfactory-and almost two hundred and fifty pounds of lovely heather honey, including a few cut combs being the end result of the hard work.

The big problem is pressing, and for this year the aim is to design myself a fairly efficient small press. The percentage losses last year were I am sure very high, and this is heartbreaking.

Finishing the season with just over nine hundred pounds was to me very satisfactory. In fact the Gales system was followed almost to the letter.

One departure is that I now have decided not only to clip all Queens, but also to mark them. Being that record cards are kept accurately on every hive, I only use one colour - white. This shows up very well and in my view saves time. Once the Queen is safely in a match box you can work that much more quickly •.

When the original idea came to me for this, I though t "Just write a short book".

What a shock was in store for me. The hours of writing and getting up at 3 am with a head full of thoughts, getting them all written down and finding on getting back to bed that something else had come to mind - up again to record these thoughts.

Chasing round in the car to see people not on the phone, and driving sixty miles to get pictures of the uncapping machine.

Writing to Brother Adam and getting his comments and advice. Then the problem of aching wrists and occasionally tearing up and. starting again.

Hope you think it's been worth it. I certainly do !